RSW

O9-AHS-018

Greenwich Observatory

300 years of astronomy

An Exhibition at the

National Maritime Museum

telling the story of Britain's
oldest scientific institution,
the Royal Observatory,
at Greenwich and
Herstmonceux

Greenwich Observatory

300 years of astronomy

edited by
Colin A. Ronan

Times Books

Acknowledgements

The publishers would like to thank all those who have helped with the production of this book, and to acknowledge the generosity of the following bodies in permitting photographs to be reprinted: the National Maritime Museum, Greenwich; the Royal Greenwich Observatory, Herstmonceux Castle, Hailsham; the Royal Society (the portrait of Nevil Maskelyne); the United States National Aeronautics and Space Administration (the space sextant, Mariner 10 and Skylab); the Cavendish Laboratory, Cambridge (the 5km radio telescope at the Mullard Radio Astronomy Laboratory); The Marconi Company Limited (Ariel 5 X-ray satellite).

First published in Great Britain in 1975 by
Times Books, the book publishing division of
Times Newspapers Limited, New Printing House Square,
London WC1X 8EZ

© Times Newspapers Limited 1975

ISBN 0 7230 0126 X

Designed by Ken Williamson
Printed and bound in Great Britain by
Cripplegate Printing Company Limited, Edenbridge, Kent

Contents

Foreword

Navigation at sea out of sight of land is a pretty hazardous business if the mariner has no means of fixing his position. It was soon discovered that observations of the sun could establish distances north or south of the equator, but distances east or west of a fixed point were much more difficult to find. Although theoretical solutions had been advanced as early as the 1530s it took over 200 years before a practical method could be devised. Three things were necessary; accurate predictions of the motions of heavenly bodies through observation; accurate measurement of time and an accurate means of measuring the positions of heavenly bodies from a ship at sea. This last requirement was achieved first when Hadley reported his invention of the reflecting quadrant to the Royal Society in 1731.

The foundation of the Royal Observatory at Greenwich provided the opportunity to make the necessary astronomical observations, although it was the German cartographer, Tobias Mayer, who worked out the first accurate Lunar Tables predicting the motion of the moon in the 1750s.

Finally, John Harrison completed his brilliant design for a Marine Chronometer, known as H4, in 1759.

That was the beginning of accurate navigation at sea, but ever since its foundation the Royal Observatory at Greenwich, and later at Herstmonceux, and the Astronomers Royal have been perfecting their services to the mariner. The fact that the Prime or Greenwich Meridian goes through the original Observatory, and is named after it, is ample tribute to its work.

The Royal Observatory has remained one of the world's centres for the determination of time and maintains its international reputation as a research centre for astronomy. This anniversary exhibition traces a long and fascinating story of progress and development. I am quite sure that all visitors will be enthralled by the displays of historic objects which have been specially brought together for this unique exhibition.

H.R.H. The Prince Philip,
Duke of Edinburgh, KG, KT

Introduction

It is very appropriate that the old Royal Observatory at Greenwich and the bulk of its surviving instruments should now form part of the National Maritime Museum, for the Observatory was established by King Charles the Second in 1675 'to find out the so-much-desired Longitude of Places for the perfecting the art of navigation'. So it was with particular pleasure that the Trustees of the Museum approved the proposal that the 300th anniversary of the founding of the Royal Observatory – Britain's oldest scientific institution – should be marked by a joint special exhibition designed by the Department of the Environment, and mounted in the Queen's House, to illustrate the world-wide influence upon the development of the oldest of the sciences, astronomy, of successive Astronomers Royal and Directors of the Royal Greenwich Observatory (as it is now known), and their staffs.

Now, of course, there is only a depleted band of astronomers at Herstmonceux Castle who can claim to have begun their careers at the Royal Observatory when it was still at Greenwich, and much though they have welcomed the dark night sky of rural Sussex that was the goal of the removal, they cannot but regret their severed links with their original home. But the Observatory still broadcasts Greenwich time, still bases the world's navigational almanacs on zero longitude at Greenwich, and is proud to be called the Royal Greenwich Observatory. Celebrating its tercentenary with the Museum through an excellent exhibition covering three hundred years of British astronomy, the Observatory looks to the future with great expectation as it enters its fourth century.

Basil Greenhill
Director, National Maritime Museum

Alan Hunter
Director, Royal Greenwich Observatory

1 Adventure in Astronomy

COLIN A. RONAN

When Charles the Second founded the Royal Observatory in Greenwich, it was no royal whim. His aim was a purely practical one – to try to solve the then intractable problem of determining longitude at sea. Technical opinion of the day had advised him that this could only be successfully achieved by astronomical sightings made on board ship, but for these methods to be a practical proposition the heavens needed re-mapping with far greater accuracy than ever before. To Charles, who had more than a passing interest in scientific matters, an observatory and an 'Astronomer Royal' were the obvious answers.

Unfortunately the money available for building the novel research institution was meagre, and although Charles commissioned Christopher Wren to design the observatory building, the funding had to come from the sale of £500's worth of spoiled gunpowder, no longer reliable enough for naval or military use, but still quite suitable for blasting. Even so the construction had to be eked out with second-hand iron, lead and bricks from a disused fort at Tilbury and a demolished gate-house at the Tower of London. Nevertheless, Wren managed to build elegantly, designing what has since become known as Flamsteed House, in a style fit 'for the Observator's habitation and a little for pompe' as he put it.

The first Astronomer Royal was John Flamsteed, a young man of twenty-nine who was granted a salary of £100 a year, no princely sum even in the seventeenth century. Moreover the Observatory possessed no instruments, and Flamsteed no assistant, except for a 'silly, surly labourer', and he had both to beg gifts of equipment and to pay out of his own pocket for some to be specially constructed. No wonder then that when he died in 1719, his widow removed all this equipment and his successor, Edmond Halley, had to refurbish the Observatory virtually from scratch. But Halley, famous now for his discovery of the comet named after him, was a very different character from Flamsteed; whereas Flamsteed suffered from ill-health, was dour and always ready to take offence, Halley was hale, sanguine and gregarious. A friendly man, he found little difficulty in persuading the Government to provide £500 for equipment. But Halley was a

FACIES SPECULÆ SEPTEN:

hard worker too, and in his twenty-two years at the Observatory managed to carry out a vast observing programme, the results of which made it quite clear that the Observatory could provide sufficiently precise measurements to make the determination of astronomical longitude a truly practical proposition.

Instruments were not all that the nascent Royal Observatory lacked: there was no policy for publication of the valuable results the Astronomers Royal were supposed to obtain. In consequence, Flamsteed amassed a quantity of observations but since he wanted to perfect them and considered them his own, he was loath to make them public. This made him the subject of severe criticism, not least from Isaac Newton, and in the end a Board of Visitors was appointed to visit and inspect the Observatory, and keep an eye on the Astronomer Royal. And as if this were not enough, Flamsteed had to suffer the indignity of having his work published in an unsatisfactory form without his permission. Halley also received similar criticism although he managed to stave off any unwarranted publication, and the question of ownership of the observations made at Greenwich was only finally resolved in 1765 when the fifth Astronomer Royal was appointed.

The publication of observations, so vital for navigation, was no less important for theoretical astronomy. Newton's criticism of Flamsteed arose because Newton wanted further detailed figures for working out his theory of the Moon's motion, and even today, in

Wren's building, known as Flamsteed House. Two of Flamsteed's long telescopes, one on the roof and one at the rear, can also be seen. From a contemporary engraving by Francis Place.

modern theoretical astronomy, precise positional determinations play a vital part. For instance, one of the present theoretical questions – whether the force of gravity varies with time – can only be finally answered by observations of time and position made with what, even now, is almost unbelievable accuracy.

The interplay between theoretical astronomy and practical navigational observing, has existed throughout the history of the Royal Observatory. Flamsteed's task was a specifically practical one according to the royal command he was given – a command he carried out with signal distinction – nevertheless he also turned his attention to theoretical questions like the paths of comets and whether the Moon possesses an atmosphere. His successor Halley followed suit, making positional measurements, particularly of the Moon, yet also publishing papers on pure astronomical research. Then when Halley was followed, as he had hoped, by his protégé James Bradley, the third Astronomer Royal took the same attitude to his commission and, indeed, is now remembered more for his pure astronomical research than for his positional measurements, excellent though they were. Bradley tried to measure the distance of a nearby star and, like other astronomers before him, failed because even his improved techniques were not refined enough to detect the minute angles involved – not until more than a century later was it possible to obtain a definite result. But Bradley's attempt was far from being unproductive. By analysing the results he did get, he discovered the first experimental evidence to prove unequivocally that the Earth really does orbit the Sun. Then, when he became Astronomer Royal, he continued these delicate measurements and in due course added further knowledge of the Earth's motion. Even Maskelyne, so absorbed with the development of astronomical tables for navigation, still found time to make measurements to allow him to compute the average density of the Earth, and observe a transit of Venus across the Sun so that the Sun's actual distance in space could be computed.

This urge to adventure into the depths of space, to probe the universe, to do more than make utilitarian measurements of star positions did not end with Maskelyne. Every Astronomer Royal has carried out research and, in the long run, this has proved a blessing rather than a curse. Astronomers of high calibre have been attracted to the staff with the result that the Royal Observatory has become a research centre of international repute. Of course, this has meant that those in ultimate control have had to take a broad view, but the Board of Ordnance did so all along, and when, in 1818, the Observatory came under the Admiralty, the attitude did not change. When Airy began his vigorous rule of the Observatory in 1835 and embarked on new areas of study, no objections were raised. Thus he could initiate regular observations in terrestrial magnetism, add meteor-

PROSPECTUS INTRA CAMERAM STELLATAM.

The 'Octagon Room' – the centre room or 'Great Room' at Flamsteed House – notable for its tall narrow windows. From a contemporary engraving by Francis Place.

ological records a few years later, and commence daily telescopic observations of the Sun without any trouble, and was even able to improve the Observatory's equipment for pure research, embarking on a programme for analysing starlight to provide evidence about the chemical nature and motions of the stars. No pure navigational research was involved in these projects; they merely added to the Observatory's lustre as a research institution.

Subsequent Astronomers Royal have continued with a judicious blend of navigational responsibility – which has included the accurate determination of time and the transmission of time signals – and pure research. In fact this double tradition became so well accepted that when Einstein published his relativity theory, it was a natural consequence that a team from Greenwich should go abroad to observe the total solar eclipse of 1919 to try to obtain experimental proof in its support.

One of the consequences of all the Royal Observatory's work, practical and theoretical, has been that first Greenwich, and now Herstmonceux, have been drawn ever more closely into the international scene. Longitude determination could not for ever remain the province of one nation, and in the nineteenth century the Observatory was involved in establishing Greenwich as the prime meridian, and in questions of the international date line and of time zones. As far as pure research is concerned, this always has had an international flavour, and ever since the transits of Venus in the 1760s, the Observatory has always been ready to co-operate in special observing programmes. Since only the northern hemisphere

Iron mural quadrant made for Edmond Halley by George Graham. With a radius of eight feet, it is shown being used as it would have been a little over two centuries ago. The large rigid framework fixed to a wall enabled the altitude of a star, or of the Moon or a planet, to be determined with considerable precision as it passed across the meridian.

skies can be observed from Greenwich or Herstmonceux, this has naturally enough led to close co-operation with observatories in the southern hemisphere, and when the Admiralty founded a Royal Observatory at the Cape of Good Hope in 1820 strong links were forged with Greenwich. The same happened when the Radcliffe Observatory moved in 1937 from Oxford to Pretoria, and now these southern hemisphere connections have become stronger still since the Cape and Radcliffe Observatories have merged with the Republic Observatory to form the new South African Astronomical Observa-

tory with Sir Richard Woolley, a past Astronomer Royal, as Director. Herstmonceux is also intimately concerned with the 150-inch Anglo-Australian telescope that H.R.H. The Prince of Wales inaugurated as recently as October 1974.

What began as a small observatory devoted to determining longitude has grown into a modern scientific research institution with international connections. Its staff may today spend as much time pushing forward the frontiers of knowledge as they do in making fundamental measurements, but nevertheless, the name of Greenwich is still, as ever, synonymous with impeccable accuracy.

2 Place in Space

DONALD H. SADLER

King Charles the Second, in 1675, commanded Flamsteed to apply himself 'with the utmost care and diligence to the rectifying of the tables of the motions of the heavens, and the places of the fixed stars, in order to find out the so much desired longitude at sea, for the perfecting of the art of navigation'.

Astronomical observations could help the mariner because the positions of the stars in the sky appear the same from all places in the same latitude, although they do this at different times according to the amount of the Earth's rotation. Thus latitude is easy to find, since it can simply be identified with a star that passes overhead, as indeed was (and still is) done by the Polynesian navigators for their island destinations. Clearly, too, the difference in longitude between two places in the same latitude may be measured by finding how far the Earth has rotated in the interval between, for instance, apparent noon at both places. Simple observation gives the local time at the ship, but the difficulty is to determine the corresponding local time at the distant place (or reference meridian).

The principle of the astronomical method of determining longitude – the technique of lunar distances – is delightfully simple. The Moon revolves round the Earth, relative to the stars, in just over 27 days and serves as the 'hand' of a natural clock, with the Sun, planets and stars as the numbers on the 'dial'. The local time of a particular configuration (for example the observed distance of the Moon from the Sun, planet or star) is compared with Greenwich Mean Time (local time at Greenwich), which can be calculated for that particular configuration from figures given in astronomical tables. The practice, however, is difficult. The Moon moves through the distance of its apparent disc in about one hour, so that to obtain longitude to half a degree its position must be observed to within one thirtieth part of its diameter – a precision that most navigators could not achieve today. Moreover, the calculations are exceedingly complex; many hundreds of methods, and tables to simplify them, were published in the eighteenth and nineteenth centuries.

This astronomical method had been described at least as early as 1514, but it still took a further 250 years for its practical realisation

The earliest known illustration of the method of lunar distances from a woodcut in *Cosmographia Petri Apian per Gemma Frisium, 1524.*

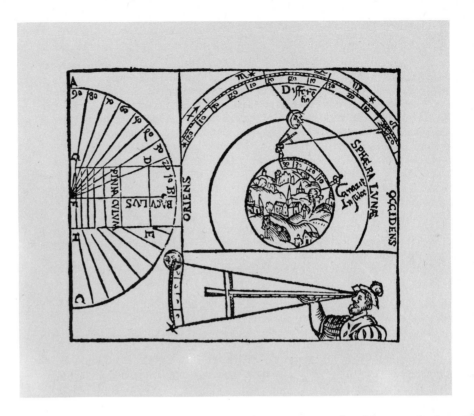

through the combination of special astronomical tables, and the invention of a good sextant. The special tables known as *The Nautical Almanac and Astronomical Ephemeris* were first published in 1767 by Nevil Maskelyne (the fifth Astronomer Royal), because only then were there enough good astronomical observations on which to base them; observations that were a tribute to the way the Astronomers Royal had fulfilled their royal command. The sextant, first described by James Hadley in 1731, was also a vital link in the chain, for with it the mariner could at last make accurate sightings of the Sun, Moon and stars at sea. With these two aids, the method of lunar distances became a practical proposition and continued to be used, in an almost unchanged manner, well into the twentieth century.

How important the method of lunar distances was may be judged from the fact that, until 1767, navigators who had 'lost their reckoning' had no alternative but to sail north or south until they reached the latitude of their destination, and then to sail due east or west. When the destination was an island, the navigator could guess wrongly and sail away from it, as Anson did as late as 1741, with the result that many of his ship's company were lost through scurvy.

There is another way to determine longitude – to use a clock to measure the interval between local time and time at a reference port like Greenwich – an obvious method that was proposed as early as 1530 by Gemma Frisius. Yet what is so simple in principle is most

Nevil Maskelyne, fifth Astronomer Royal. From a portrait by Van der Puyl, in the possession of the Royal Society.

19

"EVER since the Europeans have made long Voyages, Mathematicians have been employed in endeavouring to discover Methods, to enable them to know at all Times their precise Place at Sea. The Position of any Place on the Globe is known by its Latitude and Longitude. The Latitude of any Place is its Distance from the Equator, North or South, and is easily known by taking the Altitude of the Sun, or a Star. The Longitude of a Place is its Distance from any given Meridian, East or West, and is discovered by solving the following Problem: Knowing the Hour at the Ship, to find the Hour at the same Time at any Place whose Longitude is well known. Since 24 Hours answer to 360 Degrees of Longitude, and one Hour to 15 Degrees; if any instantaneous Appearance in the Heavens happens one Hour sooner at one Place than another, the Difference of Longitude of those Places must be 15 Degrees; that Place being to the East where it happens soonest. This Problem being of the greatest Importance to Navigation, and highly interesting to every Sailor, it may be not be amiss to shew what Attempts have been made towards the Solution of it in different Ages.

"*John Werner*, of Nuremberg, in his Annotations on the first Book of Ptolemy's Geography, printed in 1514, recommends the observing the Distance between the Moon and some Star, in order thence to determine the Longitude. *Peter Apian*, Professor of Mathematics at Ingoldstadt, in 1524, makes Mention of the same Method.

By the Assistance of Mr. *Flamstead*'s Observations, Sir *Isaac Newton* was enabled to form his curious Theory of the Moon. This incomparable Man spared no Part of that Sagacity and Industry so peculiar to himself, in settling the Epochs and other Elements of the Lunar Astronomy; but still, for want of a more continued and uninterrupted Series of Observations of the Moon, than those of Mr. *Flamstead*, the Difference of Sir *Isaac*'s Theory from the Heavens would sometimes amount at least to five Minutes.

"Dr. *Halley* employed much Time on this Subject: but not having an Instrument perfect enough to take the Distances of the Moon from the Sun or a Star at Sea, he applied himself to the Appulses and Occultations of fixed Stars.

...since our worthy Vice-President, *John Hadley*, Esq; to whom we are " highly obliged for his having perfected and brought into common Use the reflecting " Telescope, has been pleased to communicate his most ingenious Invention of an " Instrument for taking the Angles with great Certainty by Reflection, it is more " than probable that the same may be applied to taking Angles at Sea with the desired " Accuracy." This Instrument has been several Times tried at Sea, and found to answer those Expectations: the best Observers differing only one Minute in their Observations of the Sun's meridian Altitude. Considerable Improvements have lately been made in this Instrument by the Astronomer Royal...

"The greatest Reward that was offered for the Discovery of the Longitude at Sea, induced *Professor Mayer of Gottingen* to make a Set of Lunar Tables more correct than any that were before published. By the Assistance of Sir *Isaac Newton*'s Theory, his own Observations, and some furnished by *Doctor Bradley*, they were rendered so complete as not to err above one Minute in the Moon's Place.

...The bringing it into general Use was reserved for Mr. *Maskelyne* our *Astronomer Royal*. When he engaged at the Instance of the *Royal Society* to observe the Transit of Venus in 1761 at St. Helena, both in going and returning he practised this Method, and found he could always discover the Longitude within a Degree. On his Return from his Voyage, he published his Journal under the Title of The *British Mariner's Guide*, in which he suggested the Plan of a *nautical Ephemeris*, and gave a shorter and easier Method of correcting the Effects of Refraction and Parallax than any before him. On presenting his Book to the *Board of Longitude*, the *Commissioners* were so pleased with it, that they adopted his Plan; and in consequence of it ordered a *nautical Ephemeris* to be calculated, which was first published in 1767, and is now printed in advance for the Use of Sailors, ..."

Extracts from the Preface to
Tables for Correcting the Apparent Distance of the Moon and a Star . . . by Professor A. Shepherd, dated 25 March 1772.

difficult in practice. For safe navigation it is necessary to keep time to within two minutes on an Atlantic crossing, and this was long considered impossible. Only when the Yorkshire carpenter, John Harrison, made a successful marine time-keeper (the forerunner of the modern chronometer) in 1759, was it realised that the clock method was a practical one. Unfortunately the chronometer could not provide for all needs, partly owing to its rarity and expense, and partly to the necessity for checking its rate, and error, during long voyages. Lunar distances had therefore also to be used to find longitude.

Fortunately, the advent of radio time-signals provided ready checks on the error of the chronometer, thus simplifying the problem and making possible far greater accuracy. With the aid of *The Nautical Almanac*, a chronometer checked by radio time-signals and a sextant, a ship's position at sea can now be determined to within a mile or two from simple observations of the altitudes of Sun or stars that take no more than a few minutes.

In 1767, understandably enough, all the data were given in terms of Greenwich time, so that the deduced longitudes were measured from the Greenwich meridian. Other countries used other meridians but eventually, in 1884, it was agreed internationally to adopt the Greenwich meridian as the zero meridian for all purposes.

Although position at sea can now be found by radio methods using transmitters on land or in satellites, several hundred thousand almanacs are sold each year. Allowing for the smaller ships that cannot justify expensive equipment, there is still the necessity for a method that depends on the navigator, not on elaborate equipment, if only for safety's sake. Sir Francis Chichester even used his own

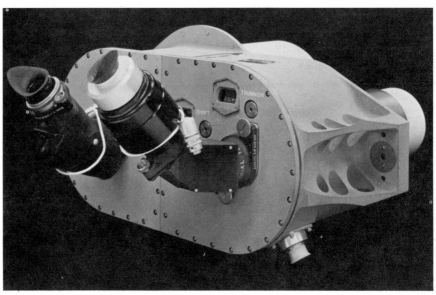

The space sextant, used for visual observations on the Apollo 8 spacecraft.

variant of lunar distances, so as to be independent of time – if necessary.

For position on land, similar astronomical methods were, and still are, used but with the advantage of larger and more stable instruments and consequently far greater precision.

Astronomical navigation was extended for use in the air from 1919 onwards, though it was not used to any considerable extent until the Second World War. The basic principles are the same as at sea, except that a vertical reference has to be used instead of the horizon. *The Air Almanac* was introduced in 1937 and, like *The Nautical Almanac*, is now a joint production with the USA. It is also used by several other countries. In the air the emphasis must be on simplicity and speed, rather than precision. However, both at sea and in the air, automatic methods of observation and calculation can now be used to provide continuous display of position to high accuracy.

Yet visual observation is not dead, even in the most advanced navigational systems. Astronomical navigation, with visual sightings, is used (though only as a back-up system) in manned space flights, with slightly different methods from those on the Earth's surface. After trials on board the Gemini flights, for example, it was used for the Apollo lunar flights and landings; the sextant sights on the Apollo 8 mission, processed on board, gave essentially the same information as the elaborate ground tracking data.

Similar methods cannot be used for the place of the Earth in space. Its position relative to the Sun has been known for centuries, with rapidly increasing precision in recent years from the analysis of radar and laser distances to the other planets. The distances to other stars are, however, so immense that only the few nearest can be directly measured, using the extremities of the Earth's orbit (300 million kilometres) as the base-line of a gigantic range-finder. But these distances, of which many have been determined at Greenwich, form the basis on which other, less direct, methods can be calibrated, leading to determinations of the distances of some of the several thousand million stars in our Galaxy, and eventually to an indication of its structure.

3 Time

HUMPHRY M. SMITH

Although the Royal Greenwich Observatory is held in high esteem among astronomers for its long series of careful visual observations, for many instrumental innovations, for wide-ranging activities in allied branches of scientific work and for valuable contributions in photographic and spectroscopic techniques, to the world at large it is renowned primarily for its unique role in timekeeping. Greenwich Mean Time (GMT) is recognised almost universally; it remains in current use in navigation, in radio communication, and as the effective reference with which national time scales and zone times must conform.

As the Earth rotates daily on its axis, we experience the regular succession of day and night. In fact, the astronomer checks the rotation of the Earth by observing the distant stars, and is then able to compare our natural clock (the Earth) with man-made clocks. The first Astronomer Royal installed two clocks, made by the outstanding clockmaker of that time, Thomas Tompion, in order to establish that the daily rotation of the Earth was sufficiently constant to justify the use of astronomical methods of navigation. He and his successors continued to observe the clock times at which certain stars, in their apparent daily motion across the sky, passed over the Greenwich meridian. These measures were used both to construct maps of the stars and also to check the Observatory clocks. Clocks and astronomical time observations have always been essential features of the work of the Royal Observatory.

An important step forward was the erection of a time ball on the roof of Flamsteed House in 1833. The ball was hoisted up the mast and dropped at exactly 1 o'clock, thus providing a visual time signal to enable ships in the Thames to set their chronometers before commencing their voyages. Some twenty years later, in 1852, a new standard clock by Charles Shepherd was installed. This employed 'galvo-magnetic' circuits to operate the time ball, and to control 'sympathetic' clocks in other parts of the Observatory. Advantage was taken of the expanding telegraph system to operate a time ball on the roof of the buildings of the Electric Telegraph Company in the Strand, and the Electric and International Telegraph Company

transmitted time signals, first to London Bridge Station, and then to the railways generally to control time balls at Deal and elsewhere.

For many years the Post Office had carried portable timekeepers in their mail coaches. In 1847 the General Post Office 'quite aware of the advantages of one uniform time system' had all their movements regulated by London time. In 1845 a Mr. Royds proposed that London time should be used throughout the railway system and that 'each company should adopt GMT as soon as the Post Office permits them to do so'. Thus, over a period of some 40 years, GMT (still often referred to as 'London Time') became generally acceptable throughout the country, although in a law case in 1858 a court ruled that local time was implied unless GMT was explicitly specified. In 1880 the need for a standard time throughout England, Scotland and Wales was formally recognised by Parliament, and GMT became the legal time.

On the other side of the Atlantic Ocean there was increasing recognition of the need for co-ordinated time, particularly as the railroads were extended further westwards from the ports on the

View of Flamsteed House showing the Time Ball in the 'dropped' position, with Greenwich and the Queen's House in the distance.

24

eastern coast. Because of the extent of the North American continent local times differed by some 5 hours between the east and west coasts, and in 1869 Charles F. Dowd put forward a proposal to establish a series of time zones, each 15° of longitude wide, which would ensure that local times differed by an exact number of hours. After some discussion on whether the time zones should be reckoned from New York or from Washington, it was agreed to press for the reference meridian to be 75° West of Greenwich. A significant consideration was that the eastern seaports had already found it convenient to maintain an exact relationship with the time of the Greenwich meridian, as the great majority of the shipping employed charts based on the Greenwich meridian, and thus used GMT for navigation. In Canada Sandford Fleming independently campaigned for a time zone system based on Greenwich, and this was eventually agreed although, very fairly, the Astronomer Royal, Airy, was reluctant to press the claims of Greenwich.

Observer using Airy's transit circle which, from 1851 to 1933, was used for time and position measurements.

The control room of the Time Department at the Royal Greenwich Observatory at Herstmonceux.

The photographic zenith tube at Herstmonceux, the most advanced type of instrument for determining time and variations in latitude.

At a congress of scientists and geodesists in Rome in 1883 it was proposed that the Greenwich meridian should be adopted as the standard for the measurement of time and longitude throughout the world, since nine-tenths of the sea charts in use were based on Greenwich, and there had been over two centuries of regular astronomical observations at the Royal Observatory. At a formal inter-Governmental Conference held at Washington in 1884, the adoption of the Greenwich meridian was agreed: 22 voted in favour, San Domingo voted against; France and Brazil abstained.

The introduction of radio time-signals marked another significant advance. The US Navy commenced transmissions in 1905, Germany in 1907 and France in 1910. In 1912 a Conference in Paris proposed the formation of an International Time Commission to secure the unification of radio time-signals, and formal statutes for an International Association of Time (now called the Bureau International de l'Heure) were drawn up the following year. In Great Britain a service of international time-signals from the Post Office Radio Station at Rugby was inaugurated in 1924, and the familiar Greenwich '6-pips' time-signals broadcast by the BBC celebrated their golden jubilee in 1974. The Post Office speaking clock was put into service in 1936.

It was not until January 1961 that the first step was taken to introduce complete synchronisation of the time signals controlled by the independent authorities in different countries, but after the UK and USA had agreed on a mutual scheme, a fully international system of co-ordination was approved in 1963.

The rate of rotation of the Earth is not uniform according to the standards of accuracy now achieved, and the development of atomic clocks has provided a basis for an independent, uniform scale of time. The internationally co-ordinated radio time-signals now conform with the scale of International Atomic Time (TAI), to which observatories and time-keeping laboratories throughout the world contribute. The Earth as a clock is at present running slow on TAI by a little over one second per year and, in order to maintain the radio time signals in approximate agreement with GMT, leap seconds have to be introduced once or twice a year.

The Royal Greenwich Observatory co-operates fully in the international work of providing the various forms of time which together meet the demanding requirements of space navigation, the tracking of artificial satellites, some advanced techniques of radio astronomy and sophisticated communication systems.

4 Observing the Universe

PHILIP S. LAURIE

When John Flamsteed was appointed Astronomer Royal, he decided that to comply with his instructions, he must first compile a catalogue of star places in the universe and make accurate predictions of the Moon's place. To this end, he obtained two fine clocks and a sextant of 7-ft radius with which to measure the angular distances between stars and to observe their passage across the meridian of Greenwich. The first astronomer to make regular use of telescopic sights for greater precision, Flamsteed later obtained a large mural arc by Sharp with which he observed the positions of the stars, Sun, Moon, and planets with an accuracy about ten times greater than any predecessor. His greatest achievement, a star catalogue, was published posthumously in 1725, and represents the Observatory's first milestone towards fulfilling its purpose.

Although these fundamental positional instruments were of prime importance, Flamsteed did have several 'gazing' telescopes, including one 60 feet long mounted on a tall mast. With these he made observations and drawings of the planets and their satellites, and of sunspots.

When he succeeded as Astronomer Royal, Edmond Halley purchased a mural quadrant of 8-ft radius for obtaining star positions. He was also a leading regular user of a transit telescope employed solely for determining time by observing the moment when stars transited across the southern meridian, but it was not until James Bradley's period of office in the mid-eighteenth century that significant improvements in observing methods took place. Bradley, probably one of the greatest observers ever, added many refinements to instrumental construction and observing techniques, producing excellent tables of the refraction or bending of light by the atmosphere, which allowed for the effects of temperature and barometric pressure on all measurements.

In 1729 he had discovered the aberration of light, and at Greenwich he remounted the telescope with which this discovery was made, a $12\frac{1}{2}$-ft zenith sector, one of the most significant of all precision instruments, and there continued the series of observations which led to his second great discovery, that of nutation, or nodding, of the Earth's axis.

Bradley also purchased a new quadrant, similar to Halley's but constructed by Bird, whose improved method of cutting divisions on such instruments enabled the Astronomer Royal and his single assistant to observe the positions of some thousands of stars with a hitherto unattainable accuracy.

Thus, by the middle of the eighteenth century, instruments and methods combined to produce results which laid the foundations of the determination of 'proper motions' (small angular motions of stars across the celestial sphere, detectable only over a long period of time), which ultimately led to knowledge of the structure of our Galaxy. Moreover, some of Bradley's observations have remained useful for more than two centuries.

Nevil Maskelyne was the first Astronomer Royal to publish his observations regularly and thereby make them available to all astronomers. He increased accuracy by applying great care to instrumental adjustments and the introduction of technical improvements, such as more efficient achromatic lenses, to his telescopes. He ordered a mural circle of new design to replace the quadrants, although it was not destined to be installed during his lifetime. But when John Pond succeeded him and substantially reorganised the Observatory's instruments, this mural circle was put to good use. Indeed Pond even added a second identical instrument, so that observations could be

The eyepiece end of James Bradley's zenith sector telescope, showing the operating controls. The eyepiece is on the left, protruding forwards and at right angles to the telescope tube, which is pivoted at its top end, some 12½ feet above the eyepiece and not shown on this engraving.

made direct by one circle and, by reflection from a mercury surface, with the other in order to eliminate errors. During this period the gift of a 4-inch aperture equatorial refracting telescope and the purchase of a 10-ft focal length reflector increased the observing capabilities to a marked degree.

The appointment of Airy in 1835 opened a new era in instrumentation of ingenuity and precision made possible by the technological advances of the day. Airy saw himself as 'not a mere Superintendent of current observations, but a Trustee for the honour of Greenwich Observatory and its general utility in the world', and proceeded to put the Observatory on a highly organised business footing. Like his predecessors, he paid great attention to the determination of the position of the Moon and undertook a massive plan for cataloguing planetary and stellar places. In 1847 he designed the Altazimuth, a special telescope capable of taking observations away from the meridian as well as on it, and thus improved lunar tables. By 1851 he had replaced the mural circles with his famous Transit Circle that still defines the zero longitude of the world, and which combined the functions of a transit telescope for time and a circle for position. Airy's Reflex Zenith Telescope paved the way for modern methods of obtaining time with even greater accuracy, and by taking advantage of the railway electric telegraph system, he installed a system of

time dissemination throughout the country. The Observatory's
largest refracting telescope in the mid-century was of but $6\frac{3}{4}$-inch
aperture but in 1855 it was replaced by the Great Equatorial, of
$12\frac{3}{4}$-inch aperture, with which further observations of the planets
were made.

The revolution in astronomy during the second half of the
nineteenth century was brought about primarily by the application
of photography and the development of the spectroscope, by which
means the light from stars could be analysed and their composition
ascertained. The whole aspect of astronomy changed at this point,
opening the way to the investigation of the physical properties of
celestial objects.

By the time of W. H. M. Christie's appointment as eighth Astronomer Royal in 1881, telescopes of greater light-grasp had become essential in order to further the study of astrophysics. A 28-inch refractor was purchased in 1894 for the observation of double stars and a 26-inch refractor designed for photographic work, coupled with a 30-inch reflector were donated by Sir Henry Thompson; all these telescopes were operating by the turn of the century.

As part of an international scheme for producing a photographic atlas and catalogue of stars, the Observatory obtained a special 13-inch refractor and by 1910, measures of some 180,000 star positions had been carried out. This type of work resulted in the development of highly accurate measuring machines, several of which were pioneered by the Observatory.

In 1911, under Frank W. Dyson, a rotatable zenith telescope floating on a bath of mercury, the Cookson Telescope, was lent by Cambridge Observatory and used for finding the variations of latitude caused by very small movements of the Earth's pole. A spectrohelioscope for studying the Sun in selected wavelengths, was obtained from Mount Wilson Observatory, California in 1929, and observation of solar flares commenced on a routine basis in order to investigate their effects on the Earth. In 1934 a gift by William Johnston Yapp added a 36-inch reflector to the Observatory's equipment; this was the largest telescope to be used at Greenwich. During this period observations were increased in scope to embrace the photographic measurement of the distances of stars (stellar parallax) as well as their magnitudes and colour.

Harold Spencer Jones became tenth Astronomer Royal in 1933, and under him a Reversible Transit Circle was brought into use as the successor to Airy's Transit Circle with which some 700,000 observations of time and position had been made and which was then approaching the end of its useful life. Spencer Jones himself made a major contribution to the determination of the distance of the Sun by using observations, made internationally in 1930–31, of the minor planet Eros. His results, published ten years later, were of an accuracy unsurpassed for the next two decades.

For many years, atmospheric pollution and the increasing number of city lights had hampered observations at Greenwich, and in 1938 Spencer Jones proposed that the Observatory should be moved to a more favourable site. Owing to the Second World War, it was not until 1946 that the Admiralty announced that Herstmonceux Castle and estate in Sussex would be the Observatory's new home. Starting in 1948, the removal took nearly ten years to complete, but included the addition of more sophisticated instrumentation among which was the Photographic Zenith Tube, now the most advanced telescope for determining time and variations of latitude. Almost twenty years later, during Sir Richard Woolley's term of office as Astronomer

Royal, control of the Observatory passed from the Admiralty to the Science Research Council. Two years later, in 1967, the Isaac Newton Telescope came into operational use; this 98-inch reflector, the largest in Western Europe, is under Science Research Council control but is managed and maintained by the Observatory for the use of all British astronomers. With moving parts nearing 100 tons, this large telescope is used for the observation of faint objects unobservable by other optical means. It is equipped with electronic image intensifiers which, when fully developed, will take the place of photographic plates, and with the Observatory's other modern equipment, will help to provide the data necessary for current investigations into the evolution of our Galaxy, and of the universe.

The Isaac Newton 98in aperture reflecting telescope at Herstmonceux.

5 The modern cosmos

WILLIAM H. McCREA, F.R.S.

The preceding articles describe past and current achievements of the Royal Greenwich Observatory in optical astronomy. We turn now to a brief review of the enormous development since the Second World War of new and unforeseen capabilities for learning about the astronomical universe by other means. Discoveries of revolutionary significance that have resulted could not have been made by optical means. Yet these discoveries provide more important work than ever for a place like the RGO. It is increasingly called upon to make its own special contributions to programmes involving observations outside its own competence. Today such co-operation involves various ways of getting extremely precise positions of the optical counterparts of radio sources and X-ray sources, monitoring their behaviour in ways it alone is equipped to do, helping in the super-accurate time-keeping needed for manoeuvring space-probes or studying pulsars – to name but a few examples.

For three centuries the Royal Observatory has observed the heavens almost exclusively by use of visible light, the only radiation we can see and, until recent times, seemingly the only radiation that could traverse our atmosphere. Until about 1940 astronomers expected progress almost solely from the use of bigger telescopes on higher mountains. We now see how things have changed dramatically in the past three decades.

Radio astronomy The coming of radio communications revealed a 'window' in our atmosphere in radio wavelengths as well as in visible light, and this led after about 1945 to the astonishing emergence of radio astronomy. Everyone knows from the press and television how it has progressed, with ever more sophisticated radio telescopes yielding ever fresh observational surprises. These include *pulsars* which are probably spinning 'neutron stars' only a few miles across; interstellar molecules radiating in unfamiliar ways; radio-emission from cold hydrogen gas, which makes it possible to trace the spiral arms of our Milky Way Galaxy; *radio-galaxies*, some of which emit more energy in radio than in optical wavelengths; *quasars*, which some astronomers think signify the birth-pangs of galaxies but more likely denote violent growing-pains, and the so-called microwave

background radiation which is interpreted as having survived from only seconds after the 'big bang' that many cosmologists regard as the creation of the universe. Even where astronomers had predicted the existence of certain of these objects, in most cases the thing they least expected was that they would first be discovered by radio means.

Radio astronomy has also seen the development of long-baseline-interferometry, as it is called, in which the use of receivers hundreds of miles apart makes possible the measurement of the angular size of radio sources no more than a few millionths of a degree across. Another development is the use of radar to measure distances and sizes of planets and satellites with unprecedented accuracy (although optical 'laser-ranging' will be even better for the nearest bodies).

Infra-red Between optical and radio there are some nine octaves of wavelength known in general terms as the infra-red part of the spectrum. While such radiation from outside the atmosphere does not normally penetrate to the Earth's surface, some may reach places at high altitude when the atmosphere is specially dry. It can then be focused by a reflecting telescope, but it can be studied only by devices evolved in recent years. For several years there have been some such installations in the US, and UK astronomers have operated some in Tenerife and are preparing to operate in Hawaii. With aid

The 5km radio telescope at the Mullard Radio Astronomy Observatory, Cambridge.

also from equipment in aircraft and space vehicles, there has thus emerged the subject of infra-red astronomy. Again the results are surprising. Within our Galaxy there are sources of considerable amounts of infra-red emission; many are commonly believed to be regions where stars are still condensing out of diffuse material; some astronomers think others may be highly luminous stars overlaid by dense dust-clouds. Certain other galaxies and certain quasars emit more energy in the infra-red than in all other wavelengths.

Space astronomy Radiation in the far ultra-violet and beyond, coming from outside the atmosphere, cannot penetrate to any ground-based observatory; detecting devices have to be sent aloft on rockets and satellites. In particular, the US National Aeronautics and Space Administration has since 1962 operated several 'orbiting observatories'; it has recently operated *Skylab* as the first manned space observatory, mainly for solar work so far as astronomy is concerned, and it will soon have something even more sophisticated. Here we may mention only a couple of uses of these unexampled facilities.

Space astronomy finds that X-rays are emitted by active regions of the Sun's atmosphere, by discrete sources within the Galaxy – one being in the Crab Nebula and some being double stars in which

The Earth forms a dramatic backdrop for the orbiting Skylab space station in this view taken by the third visiting crew on departure. At the far end is the Apollo telescope mount with its paddle-wheel power panels.

Scientific equipment of Ariel 5, the most recent satellite in the British collaborative programme with NASA. Ariel 5 was launched on 15 October 1974, and is an advanced X-ray satellite designed to carry out the most comprehensive investigation yet attempted into X-ray sources, including phenomena possibly associated with black holes.

one member is highly condensed – by some galaxies and quasars which emit more energy as X-rays than in any other way, and by some regions in clusters of galaxies; there is also a weak X-ray background in the cosmos as a whole. None of these discoveries was predicted, and few have yet received agreed explanations – a phrase that becomes a refrain in this narrative.

The most energetic quanta of radiant energy known to physics are *gamma-rays* that are emitted solely in nuclear processes. Only in the past couple of years have space astronomers succeeded in positively detecting such rays outside the Earth. Unexpectedly, they arrive apparently in surges, and no one yet knows where they come from. They may be counted upon ultimately to give important insight into 'high-energy astrophysics'.

Astronomy has thus covered the whole gamut of radiant energy.

While it is only beginning to gather all the new knowledge that must result, it is manifest that the present generation has experienced one of the great epochal phases of scientific endeavour.

Corpuscular astronomy Cosmic rays are nuclei of atoms, predominantly hydrogen, that reach the top of the atmosphere with enormous energy. Their study by instruments borne aloft in various vehicles, and by instruments on the ground measuring secondary effects, is now a recognized branch of astronomy. Some of the less energetic of the particles come from solar eruptions, but the general view is that the rest come from supernova explosions in our Galaxy, and probably other galaxies too. So they tell us about the composition of these remote bodies, and about matter and magnetic fields in intervening space. For one thing, they strongly indicate that the observable universe is all composed of our sort of matter, not half-and-half matter and anti-matter as some astronomers had conjectured.

While cosmic rays probably come from short-lived stellar explosions, according to standard astrophysics any normal star like the Sun continually emits a great flux of particles known as *neutrinos*, produced by the nuclear reactions in the central regions that keep the star shining. Neutrinos pass almost freely through the Sun and the Earth and they are fantastically hard to detect. However, physicists who have worked out a means of doing so have constructed the most paradoxical of all observatories; it is for 'looking' into the centre of the Sun and it has to be operated in the deepest mine available (so that the overlying ground traps unwanted particles). Again the result is a surprise – currently the most embarrassing puzzle in astronomy – for *no* solar neutrinos have been found!

Space probes Were the Earth to lose its atmosphere, but not its astronomers, all the foregoing observations could be made from the ground; orbiting observatories and so on would be irrelevant. But there would remain the same need as at present for the other sort of 'space-astronomy' using space-probes, i.e. going or sending instruments to members of the Solar System to view features that could never be seen from afar. The Apollo, Mariner and Pioneer missions from the USA and the Luna and Venera from the USSR have in a few years taught us more about the constitution of the System than had been learned in all past ages.

Gravitation Were our planet forever cloud-bound, we could nevertheless discover the existence of our Moon by observing the ocean tides. It may not, therefore, be surprising that during recent years 'observatories' have been established to try to observe the heavens using gravity – not the steady pull of bodies as tidal force but *gravitational waves* that are inferred to be generated by gravitationally interacting bodies, including, maybe, masses falling into *black holes*. Many astronomers believe that this will be the most exciting next stage, but results so far are controversial.

Mariner 10 spacecraft showing solar panels, radio 'dish' aerial, cameras and other equipment. This spacecraft has made close approaches to Venus and Mercury.

During the past 30 years man has newly opened upon the universe many eyes that during the rest of the 300 years of the RGO he never knew he possessed. With every eye he has seen things he had not known to exist. But they all form one universe and astronomy remains one science. As indicated at the outset, the optical observatory, particularly a comprehensive observatory like the RGO, gains in importance with every fresh discovery by other means. For when anything new is discovered by any means the natural urge is to try to find something to do with it that can be looked at or photographed. Inevitably the optical observatory is therefore the place where all other observations tend to get linked together, and hopefully made comprehensible. There is plenty to keep the RGO busy for another three centuries.

Glossary

Achromatic	Colour free. An achromatic lens is a compound lens that brings the light of more than one colour to focus at the same point, thus largely eliminating the coloured fringes seen through an ordinary simple lens.
Anti-matter	Matter composed of elementary particles similar to those of ordinary matter, but with opposite electrical charge.
Astrophysics	The study of the physical nature and behaviour of celestial bodies.
Big bang	The explosive start to the universe some ten thousand million years ago, inferred from the more generally accepted theories of cosmology.
Binary star	A pair of stars in orbit around one another.
Black-hole	A mass of material compressed to such density that no light or any other form of energy can escape from its gravitational pull.
Ephemeris	A table giving the computed places in the sky of celestial bodies.
Galaxy	An agglomeration of anything between some thousand million and hundred thousand million stars, together with dust and gas, forming a recognisable unit with its own evolution and movement. Our own Milky Way system is just such a galaxy.
High energy astrophysics	The study of astrophysics associated with atomic particles having very high energies (sometimes called 'relativistic' energies).
Infra-red	Radiation of wavelength greater than visible light, and less than radio wavelengths.
Interferometer	An optical or radio device for measuring very small angles by exploiting the wave nature of the radiation. In radio astronomy, an interferometer can be formed by two or more radio telescopes linked together electronically and observing the same radio source.
Laser ranging	Measuring distance by a method similar to radar, but using an intense narrow beam of light instead of radio waves.

Magnetic field	The region surrounding a magnetised body, and in which magnetic forces can be detected.
Microwaves	Very short wavelength radio waves, a few millimetres or centimetres in length.
Molecule	A combination of atoms forming a unit of chemical structure.
Neutron star	A star of density so great that most of its material is concentrated in the form of particles known as neutrons, rather than distended normally in the form of atoms.
Octave	The interval between a frequency (wavelength) and double that frequency.
Optical wavelengths	Visible light, ranging from violet (shorter wavelengths) to red (longer wavelengths).
Parallax	A measure of distance. To determine distances of nearby stars, observations are made of them at six-monthly intervals, when the Earth is at the opposite ends of its orbit. Such observations, made at two different points in space, will cause the star to appear to shift across the background of more distant stars. This shift is measured as an angle, the size of which depends on the star's distance. Half this angle is known as the parallax.
Quanta	The plural of quantum. Atoms and atomic systems in general lose or gain energy in discrete amounts known as quanta.
Quasar	A celestial body that looks like a star, yet is not one. Such a quasi-stellar object emits about 100 times as much radiation as a galaxy does, but its precise nature is at present undetermined.
Radio source	A celestial object that emits detectable radio radiation.
Reflector	A telescope which has a concave mirror to gather light and bring it to a focus.
Refractor	A telescope in which the light is gathered and brought to a focus by a lens, not a mirror.
Sextant	An instrument with a graduated scale of one sixth of the circumference of a circle. Used for measuring angular distance between a celestial object and the horizon or another celestial object.
Solar flare	A bright eruption of hot gas on the Sun.
Solar system	The system of bodies – planets, minor planets, satellites, meteors and comets – that orbit the Sun.
Supernova	An exploding, disintegrating star.
Total solar eclipse	Occurs when the Moon is seen to obliterate the Sun's disk. At this time it is possible to pick out stars and planets in the sky.
X-rays	Very penetrating radiation lying beyond the ultraviolet and with a wavelength some ten thousand times less than that of visible light.

Exhibition Organising Committee

National Maritime Museum

Basil Greenhill CMG   Director
David Waters   Deputy Director & Head of Division of Navigation & Astronomy

Derek Howse   Head of Department of Astronomy
Sue Gaston   Research Assistant
Angela Molyneux   Assistant Designer
Reginald Barrow CVO   Project Manager

Royal Greenwich Observatory, Herstmonceux

Alan Hunter   Director
David Thomas   Assistant to the Director
Philip Laurie   Archivist

Department of the Environment

John Pound   Designer

Astronomers Royal and Directors of the Royal Greenwich Observatory

John Flamsteed	First Astronomer Royal 1675–1719
Edmond Halley	Second Astronomer Royal 1720–1742
James Bradley	Third Astronomer Royal 1742–1762
Nathaniel Bliss	Fourth Astronomer Royal 1762–1764
Nevil Maskelyne	Fifth Astronomer Royal 1765–1811
John Pond	Sixth Astronomer Royal 1811–1835
George Airy	Seventh Astronomer Royal 1835–1881
William Christie	Eighth Astronomer Royal 1881–1910
Frank Dyson	Ninth Astronomer Royal 1910–1933
Harold Spencer Jones	Tenth Astronomer Royal 1933–1955
Richard v.d. R. Woolley	Eleventh Astronomer Royal 1956–1971
Martin Ryle	Twelfth Astronomer Royal 1971–
E. Margaret Burbidge	First Director of the Royal Greenwich Observatory 1972–1973
Alan Hunter	Second Director of the Royal Greenwich Observatory 1973–

Guide to the Exhibition

The National Maritime Museum is extremely grateful to all those who have lent objects for this exhibition. Their names are given under the appropriate entry in the list below. Where no name is shown, the object is the property of the Trustees of the National Maritime Museum.

ORO = Old Royal Observatory, National Maritime Museum, Greenwich

RGO = Royal Greenwich Observatory, Herstmonceux

Area 1
Entrance to Great Hall

Introduction

1 1675

a Ship of 90 guns; Admiralty model, scale 1:48; *c.*1675. Ref: 1675–1.

b Replica of Isaac Newton's reflecting telescope. The original was made with his own hands in 1671 and is in the possession of the Royal Society.

Lent by the Royal Society.

c Interior view of the Isaac Newton Telescope. RGO photograph.

2 1975

a Model of 98in Isaac Newton reflecting telescope (INT), Herstmonceux, 1967.

Property of RGO.

b Model of Orbiting Astronomical Observatory (OAO-3), launched 21 August 1972.

Lent by National Aeronautics and Space Administration, Washington DC.

3 Louise de Kéroualle, Duchess of Portsmouth – a favourite of Charles II

a *The Duchess of Portsmouth*; oil painting by P. Mignard, 1682. 120.7×95.3cm/$47\frac{1}{2} \times 37\frac{1}{2}$in. In 1674 a Frenchman, friend of the Duchess, claimed to have discovered a sure method of finding longitude at sea. The claim proved unfounded but the King was persuaded that a Royal Observatory ought to be set up.

Lent by the National Portrait Gallery.

4 The instigator and benefactor of the Royal Observatory – Sir Jonas Moore FRS
Surveyor General of the Ordnance

a *Effigies Ionae Moore Matheseos Professoris Ætat: Suae 45 Anº: Dni: 1660*; engraving from Moore's posthumous *A New Systeme of Mathematicks . . .*, London 1681.

Property of RGO.

b *A Mapp or Description of the River of Thames . . . made by Jonas Moore Gent . . . 1662.* Pen, watercolour and gold paint on vellum. 64.7×223.5cm/$25\frac{1}{2} \times 88$in. The inset view of Greenwich, drawn by Hollar, shows Greenwich Castle on the site later occupied by the Royal Observatory.

Lent by the London Museum with the agreement of the Public Records Office.

5 The architect – Sir Christopher Wren, FRS
King's Surveyor General and one-time Professor of Astronomy

a *Sir Christopher Wren*; oil painting by J. B. Closterman, 128.3×105.4cm/$50\frac{1}{2} \times 41\frac{1}{2}$in. The Observatory was built to Wren's design in 1675 'for the observator's habitation and a little for pompe'.

Lent by the Royal Society.

6 The Royal Observatory about 1680

a *Greenwich Observatory from Croom's Hill*; oil painting by unknown artist *c.* 1680. 101.6×167.6cm/40×66in. Ref: 46–260.

Catalogue cover illustration.

7 The Founder – King Charles II

a *Charles II*; oil painting by J. M. Wright, *c.*1680.

Lent by The Department of the Environment.

8 Navigation; astronomy; time

a Globes, celestial and terrestrial, by John Senex, London, before 1740. 68cm/$26\frac{3}{4}$in diam. Ref: G138/9.

Colour plates 4 and 5

b Abraham Sharp's double refracting telescope, on an equatorial mounting, *c*.1700.
Ref: OO/R.22.
Sharp (1653–1742) was assistant to Flamsteed at Greenwich in 1684 and 1688–9.

Property of the Yorkshire Museum, York.

Colour plate 2

c John Harrison's first marine timekeeper, 1735.
Ref: Ch.35.
The forerunner of the watch which won the Longitude Prize of £20,000 in 1773, exhibited today in Navigation Room of the National Maritime Museum.

Property of the Ministry of Defence (Navy).

Colour plate 1

9 The age of discovery

a *Portuguese Carracks*; oil on panel, school of Patinir, *c*.1525.
78.7 × 144.8cm/31 × 57in.
Ref: 35–22.
Probably the best contemporary representation of the first generation of these ocean-going merchantmen, whose ocean voyages pointed the need for methods of finding longitude at sea.

Colour plate 3

10 The observatory founded, 1675

a Contemporary copy of Royal Warrant for building Greenwich Observatory, June 22 1675.

Property of RGO (MS 35/8).

The navigational problem

11 Navigation in 1675

a Frontispiece of J. Seller, *Practical Navigation*, London 1669.
Photograph.

b Woodcut of ship, detail from title page of John Tapp, *The Seamans Kalendar*, London 1617.
Photograph.

12 Navigational equipment of 1675

a–d Cross-staff, backstaff, 24in Gunter's scale and 12in Gunter's scale; ivory presentation set by Thomas Tuttel by Charing Cross, London *c*.1700.
Ref: S117–8, CI/GS 1–2.

Colour plate 6

e Gunter's sector, ivory, signed *Tho: Tuttel Charing + Londini fecit 1699*.
Ref: CI/S23.

f Sandglass, half-hour; turned ebony and ivory, 17th century.
Ref: H33.

Colour plate 6

g Universal equinoctial ring dial; brass, 13.5cm/$5\frac{5}{16}$in diameter, inscribed *H. Bedford by Holborn Conduit* [London] *c*.1680.
Ref: D295.

Colour plate 7

h Lodestone mounted in brass, 17.0 × 10.2 × 12.7cm/$6\frac{3}{4}$ × 4 × 5in, unknown maker, *c*.1680–1700.
Ref: M.7.

i Dividers, brass with steel points 29.2cm/ $11\frac{1}{2}$in long, inscribed *Capt In° Kempthorne*, *c*.1650
Ref: DI/D1.

j Manuscript chart of North Atlantic; pen and watercolour on vellum, on four hinged wood boards; inscribed *Made by Nicholas Comberford Dwelling Neare To the West End of the Schoole House at the signe of the Platt in Redcliffe Anno 1650*.
Ref: G213: 2/2.

k J. Seller, *Practical Navigation*, 2nd edition, London 1672.

13 Shipwreck!

a *English Men of War wrecked on a rocky coast,* oil painting by Willem van de Velde the Younger (1633–1707).
124.5 × 179.1cm/49 × 70in.
Ref: 69–92.

14 The nature of the problem

a Detail from frontispiece of W. Blaeu, *The Light of Navigation,* Amsterdam 1620.
Photograph.

b *The Figure of the Quadrant,* from J. Seller, *Practical Navigation,* 2nd edition, London 1672.
Photograph.

Area 2

First West Room
1675–1720

1 John Flamsteed, first Astronomer Royal, 1675–1719

a *John Flamsteed;* detail from engraving by Geo Vertue (1721) after J. Gibson (1712). Photograph.

2 Measuring the Earth's rotation

a The clocks in the Great Room about 1676, detail from etching *Prospectus intra Cameram Stellatam*, by F. Place after R. Thacker. The clock marked *C* on the etching has not survived. Photograph.

b *Tho: Tompion Automatopoeus;* mezzotint by I. Smith after G. Kneller (1713). 29.2×25.1cm/$11\frac{1}{2} \times 9\frac{7}{8}$in.

c A Year Clock inscribed *Motus annuus. Sr Jonas Moore caused this movement with great care to be thus made A^0 1676 by Tho Tompion.* Originally one of a pair mounted behind the wainscot in the Great Room at Greenwich with a 2-second pendulum hung above the movement (see 2a above), this clock was converted to operate with a conventional 1-second pendulum and placed in the present case some time in the 18th century. The second clock of the pair is today in the British Museum.

Lent by the Earl of Leicester.

3 The fruits of fifty years' labour

a J. Flamsteed, *Historia Coelestis*, London 1712.
Sometimes called Halley's 'pirated' edition, in one volume with a Preface and star catalogue (which Flamsteed called 'spurious') by Edmond Halley.

Property of RGO.

b Letter dated 14 November 1705 from Isaac Newton to Flamsteed about the publication of *Historia Coelestis*. RGO Ms Vol 35 folio 31.

c J. Flamsteed, *Historia Coelestis Britannica*, volume 3, London 1725.
Published posthumously in three volumes, this contained Flamsteed's British Catalogue of star positions.

Property of RGO.

4 John Flamsteed

Oil painting signed *T. Gibson fec^t 1712*. 120.7×98.4cm/$47\frac{1}{2} \times 38\frac{3}{4}$in.

Lent by the Bodleian Library, Oxford No. 212.

5 Correspondence with foreign astronomers

a Johannes Hevelius to Flamsteed, Danzig, 24 April 1679.

b James Pound to Flamsteed, Batavia, 14 March 1702.

c Gottfried Leibnitz to Flamsteed, Berlin, 17 April 1703.

d Henry Stanyan to Flamsteed, Berne, 23 October 1706.

e Thomas Brattle to Flamsteed, Boston in New England, 31 May 1707.

6 Flamsteed's star atlas

a J. Flamsteed, *Atlas Coelestis . . . ,* London 1729.

7 How far are the stars?

a Object glass by Pierre Borel of Paris, c.1677, inscribed:
Flamsteed's O.G. 90ft fo^s. presented by Jm^s Hodgson Esq Nov 3 1737 Royal Society No. 25.
Used by Flamsteed in the unsuccessful well-telescope with which he hoped to measure the parallax* of the star *Gamma Draconis*.

Property of the Royal Society.

Lent by the Science Museum (Inv. No. 1932–466).

*See glossary on page 42.

b Flamsteed's well telescope, detail from etching *Puteus 100 pedum ad Parallaxes Terrae observandes, p^r paratus* by Francis Place after Robert Thacker, c.1676.
Photograph.

c John Flamsteed, cast of ivory relief by David le Marchand, the original of which is the property of the RGO, inscribed on the front *D.L.M.* and on the back *IOH: FLAMSTEEDIVS Math. Reg. Ætat An. 73 completo 1719* [the year of Flamsteed's death].

8 The tides at London Bridge

a London Bridge about 1700; detail from the engraving *Prospect of the City of London* by unknown artist, from *Le Nouveau Théâtre de Grande Bretagne*, Vol. 2, 1st ed., London 1713.
Guildhall Library photograph.

9 Coffee-houses in 17th century London

a *Interior of a Coffee house*; watercolour by unknown artist, inscribed *AS.1668 c.1705*.
British Museum Photograph.

10 Sir Isaac Newton, mathematician and President of the Royal Society

a Portrait; oil painting by John Vanderbank 1726.
125.8×101.0cm$/49\frac{1}{2} \times 39\frac{3}{4}$in.

Lent by the Master and Fellows of Trinity College, Cambridge.

11 Newton's law of gravitation

a Putting a projectile into orbit. From English translation of Newton's *Principia, System of the World*, London 1728.

Ronan Picture Library photograph.

12 A quarrel

a Newton's receipt for Flamsteed's lunar observations. RGO MSS 6,f181.
Photograph.

13 A Board of Visitors appointed, 1710

a Royal Warrant signed by Queen Anne, appointing the President and nominated members of Council of the Royal Society as Visitors to Royal Observatory.

Lent by the Royal Society (MSS 372/96).

14 The Longitude Act, 1714

a *An Act for providing a Publick Reward for such Person or Persons as shall Discover the Longitude at Sea* (13 Anne, Cap. 15), 1714.
Photograph.

b Portrait of William Whiston.
Photograph by courtesy of Sir John Conant, Bt.

c Title page and two maps showing lines of equal dip from *The Longitude and Latitude found by the Inclinatory or Dipping Needle*, by William Whiston, London, 2nd ed., 1721.

15 Flamsteed's copy of Newton's *Principia*

a I. Newton, *Philosophiae Naturalis Principia Mathematica*, London 1687.
Flamsteed's own copy.

Lent by the Royal Society.

16 Hadley's reflecting quadrant

a 20in Hadley reflecting quadrant, mahogany with boxwood scale, inscribed *Made by Geo Adams in Fleet Street, London, c.1750* (ref. 5.5).

b *John Hadley Esqr VPRS.* Portrait; oil painting attributed to Bartholomew Dandridge, c.1720–30.
41.9×32.4cm$/16\frac{1}{2} \times 12\frac{3}{4}$in.

17 Edmond Halley, second Astronomer Royal, 1720–42

a Portrait, from engraving by Geo Vertue after R. Philips.
Science Museum photograph.

18 Halley – the Southern Tycho, 1677–8

a Pair of celestial planispheres, *The Right Ascensions and Declinations of the Principal Fixed Stars in Both Hemispheres to ye year 1678* after Edmond Halley, each 53 × 50cm/21 × 19¾in.
Ref. G200: 1/3 A and B.

b *St Helena*, etching by unknown artist, *c.*1670.
Photograph.

19 Halley and Peter the Great, 1698

a *Petrus Primus Russorum Imperator*, etching by J. Houbraken.
Photograph.

b *Plan of Deptford, showing Sayes Court* (detail), from H. B. Wheatley (ed), *Diary of John Evelyn*, II, London 1906.
Photograph.

20 Halley the Navigator, 1698–1701

a Edmond Halley, *A New and Correct Sea Chart of the Whole World Shewing Variations of the Compass as they were found in the year MDCC* (left half only), engraving on paper, hand-coloured, Mount and Page, London *c.*1710–30.
Ref. G201: 1/1A.

b *An English Flute [or Pink] off Satalia* from a grisaille by Jan Peeters.
Photograph.

21 Halley's comet

a Halley's comet in 1066; part of Bayeux Tapestry.
Photograph by arrangement with Maison Combier, 4 rue Agut, 71 Mâcon, France.

22 Halley as Astronomer Royal, 1720–42

a A page from Halley's observation book for August–September, 1730. (RGO MSS Vol. 79 f.95).
Photograph.

23 DR E^D HALLEY AGED 80

Oil painting by Michael Dahl, 1736.
127.0 × 101.6cm/50 × 40in.

Lent by the Royal Society.

Area 3
Second West Room
1742–1835

1 James Bradley, third Astronomer Royal, 1742–62

a James Bradley; detail from engraving after Jonathan Richardson.
Ronan Picture Library photograph.

2 James Bradley

Oil painting by Thomas Hudson, 73.7 × 62.3cm/29 × 24½in.

Lent by the Bodleian Library, Oxford (No. 258).

3 The discovery of the aberration of light, 1727

a Bradley's model illustrating the theory of the aberration of light; wood, c.1730–40. This model was devised and used for demonstration purposes by Bradley himself.

Lent by the Science Museum (Inv. No. 1876–1029).

4 The marine sextant

a 18in Hadley reflecting quadrant; brass scale, English c.1770.
Ref. S10.

Colour plate 8

b 18¼in sextant, brass, fitted with wooden belt-pole and frame, signed *J. Bird, London, c.1770.*
Ref. S.225.

c 8in pillar frame sextant; brass with wooden box, signed *Ramsden London 1421,* c.1795.
Ref. S31.

Colour plate 10

5 Calendar reform, 1752

a Lord Chesterfield, after W. Hoare, c.1742.
National Portrait Gallery photograph.

b 'Give us our eleven days', detail from the engraving *The Election, Plate 1 – Humours of an election entertainment,* by T. E. Nicholson after W. Hogarth.
Photograph.

6 Tobias Mayer's lunar and solar tables

a *Tobias Mayer,* engraving by Westermayr after Kaltenhejer.
Photograph.

b Extracts from Tobias Mayer, *Tabulae Motuum Solis et Lunae . . .,* London 1770.
Photograph.

7 Accuracy of observations

a Bradley's quadrant observations August–September 1752, from his *Astronomical Observations . . . Greenwich . . . ,* II, Oxford 1798, page 39.
Photograph.

This page covers the date when Britain changed from the Julian to the Gregorian calendar – when September 3 became 14 September 1752.

8 Bradley's own achromatic telescope

Overall length 221cm/7ft 3in, aperture 7.3cm/2⅞in with ten-sided mahogany body tube and short brass draw tube signed 'Dollond, London'.
Bradley died only four years after Dollond first described the achromatic telescope. This must therefore be one of the earliest of its type still in existence.

Lent by the Museum of the History of Science, Oxford.

9 The achromatic telescope

a Diorama of John Dollond's workshop; modern.

Property of RGO.

b *John Dollond* (1706–61), from an engraving by John Possel White.
Photograph.
John Dollond first described the achromatic telescope in 1758.

c *Peter Dollond* (1730–1820), from an oil painting by an unknown artist in the possession of Dollond & Aitchison Group Ltd.
Photograph.

d Receipted invoice for 5ft achromatic telescope, etc. signed by Peter Dollond on 3 February 1764.
Dollond & Aitchison Group photograph.

10 Nathaniel Bliss, fourth Astronomer Royal, 1762–4

a Nathaniel Bliss, from an engraving on an old pewter flagon, inscribed: *This sure is Bliss, if Bliss on Earth there be,* reproduced in E. W. Maunder, *The Royal Observatory, Greenwich*, London 1900, page 83.
Photograph.

11 Nathaniel Bliss

Oil painting by unknown artist *c.*1755, 78.7 × 66.0cm/31 × 26in.

Lent by M. L. Dix-Hamilton Esq.

12 Nevil Maskelyne, fifth Astronomer Royal, 1765–1811

a Nevil Maskelyne, detail of engraving after Van der Puyl, from *The Gallery of Portraits*, Vol. VI, London 1836.

13 The Nautical Almanac

a N. Maskelyne (ed), *The Nautical Almanac and Astronomical Ephemeris for the Year 1767*, London 1766.

b *The Nautical Almanac for the year 1975*, London and Washington DC 1973.

c N. Maskelyne (ed), *Tables Requisite to be Used with the Astronomical and Nautical Ephemeris*, 1st ed. London 1766.

d N. Maskelyne, *The British Mariners' Guide*, London 1763.
This copy bears the signature of Margaret Maskelyne, daughter of the Astronomer Royal.

c and **d** *lent anonymously.*

14 Nevil Maskelyne – personal items

a *Nevil Maskelyne*, Astronomer Royal; pastel by John Russell, 1804, 58.4 × 43.8cm/23 × 17$\frac{1}{4}$in.

b *Sophia Maskelyne*, wife of the Astronomer Royal; pastel by John Russell, 1804, 58.4 × 43.8cm/23 × 17$\frac{1}{4}$in.

c Maskelyne's observing suit; trousers, jacket and topcoat; quilted Indian silk; *c.*1765.
Specially made in India to the order of Baron Clive, Governor of Bengal and Nevil Maskelyne's brother-in-law.

d Pair of 18in globes, by W. and S. Jones; terrestrial dedicated to Sir J. Banks, celestial to Nevil Maskelyne, *c.*1806.
Purchased by Margaret, Lady Clive, younger sister to Nevil Maskelyne.

All these items lent anonymously.

15 Science and exploration

a Letter from Benjamin Franklin (10 March 1779) 'to all captains and Commanders of armed ships acting by Commission from the Congress of the United States of America now in War with Great Britain', requesting that Captain Cook's ships – whose return from the third voyage was imminent – be not molested despite the state of war which existed.

Lent by the American Philosophical Society (F$\frac{B}{85}$394).

b Letters from M. de Sartines (13 April 1778) and Duc de Croÿ (18 February 1779) issuing similar instructions to the French Navy and privateers. Archives Nationales Marines B^{4}315, 1er dossier.
Photograph by courtesy of SIRCO–France.

c Regulator clock by Thomas Earnshaw, London, purchased by Board of Longitude, 1791; used on voyages of exploration of Vancouver (NW America 1791–4) and Flinders (Australia 1801–2); also for Transit of Venus (Hawaii 1874).

Lent by Mark Dineley Esq.

16 *The Distinguished Men of Science of Great Britain living in the years 1807–8.*
Engraving by W. Walker and G. Zobel after J. Gilbert.
Original drawing in National Portrait Gallery.
Published 4 June 1862.

17 The transits of Venus, 1761 and 1769

a 12in portable astronomical quadrant, brass, inscribed *John Bird London, c.1768.*
Made for the transit of 1769, and said to have been taken by Captain Cook on at least one of his voyages, 1768–80.

Property of the Science Museum (Inv. No. 1876–572).

Colour plate 9

b 24in Gregorian reflecting telescope; brass, inscribed

James Short London $\frac{51}{1267} = 24, 1763.$

Similar to the telescopes used by most British observers at the transits of 1761 and 1769.

Property of the Whipple Science Museum, Cambridge (No. 1076).

Colour Plate 11

c Astronomical Regulator Clock on wooden tripod, inscribed *Royal Society No. 35. John Shelton London,* 1768.
Probably used at the North Cape, Norway, for the transit of 1769, this was almost certainly one of the clocks which accompanied Captain Cook on his second and third voyages, 1772–80.

Property of the Royal Society.

Colour plate 12

d Cook's Tent observatory, 1776; engraving by J. Basire; Plate II in W. Wales, *Astronomical observations . . . Resolution and Adventure . . . 1772–5,* London 1777.
Photograph.

This portable observatory was designed by John Smeaton, the famous civil engineer.

18 The development of the marine chronometer

a 2-day chronometer, signed: *Pennington Pendleton and others: for the son of the Inventor* [Thos. Mudge] *No. 12, 1796.*
Ref: Ch154.

Property of the Ministry of Defence (Navy).

b Longitude-by-chronometer calculations; from J. Weddel, *A Voyage towards the South Pole, 1822–24,* London 1827, page 242.
Photograph.

19 John Pond, sixth Astronomer Royal, 1811–35

a *Mr Pond*; engraving by Benjamin Smith after Thomas Parkinson.
Photograph.

20 The Admiralty takes control

a *Account of the rate of the several chronometers at the Royal Observatory, for July 1825*; pamphlet, Greenwich 1825.
Photograph.

21 The time ball

a Operation of the time ball; engravings from *The Illustrated London Almanack,* 1845, page 28.
Photograph.

b Notice to mariners of 28 October 1833, announcing the establishment of the Greenwich time ball; from *The Nautical Magazine,* London 1833, page 680.
Photograph.

Area 4
North Side
of Great Hall

1835–1950

1 George Biddell Airy, seventh Astronomer Royal, 1835–81

a George B. Airy, lithograph by unknown artist, c.1850.
Photograph.

2 The spread of Greenwich Time

a 'The Timekeeper of the Country' from *The Graphic*, August 8, 1885.

b 'Electric time-ball and clock, West Strand', London, 1852; from The *Illustrated London News*, September 11 1852.
Photograph.

c 'The Astronomer Royal–Mr Punch's fancy portrait No. 134,' from *Punch*, 1883, p.214.
Photograph by courtesy of *Punch*.

3 Navigational equipment of 1875

a Dead reckoning – steering compass, the card inscribed *E. I. Dent . . . London*, c.1850.
Ref: C.130.
Recording log by Massey of London, c.1810.
Ref: Lg.19.

Colour plate 14

b Nautical astronomy – 7¼in sextant by Reynolds and Wiggins of London, c.1850.
Ref: S.53.
Marine chronometer by Charles Frodsham of London, No. 2241, c.1850.
Ref: Ch.46.
The Nautical Almanac and Astronomical Ephemeris for the year 1851, London 1847; J. W. Norie, *A Complete set of Nautical Tables*, 11th edition, London 1845.

Colour Plate 13

c Chartwork – Admiralty Chart No. 1689, *Madeira: Funchal Bay*, surveyed by Captain A. T. E. Vidal 1843, London 1845.
Ref: G214:24/2.
Martin White, *Sailing Directions for the English Channel*, 3rd edition, Hydrographic Dept., London 1846.
24in rolling parallel ruler, brass, by Crichton of London, c.1820.
Ref: DI/RP.23.
Dividers, brass, by Thomas Jones of Charing Cross, London, c.1840.
Ref: DI/ST.14 part.
Black lead pencil, by E. Wolff & Son, c.1840.

4 George Airy

Oil painting by James Pardon, 1833.
139.7 × 109.2cm/55 × 43in.

Lent by the Institute of Astronomy, Cambridge.

5 Magnetism and meteorology

a Vertical force magnetograph, 1841.

b Dip circle inscribed: *Troughton & Simms London 1860*.

c Photographic recording of magnetic elements and barometer, 1870.
Photographs.

Property of RGO.

6 Nineteenth-century technical developments – spectroscopy, photography and solar physics

a 6½in objective prism by A. Hilger c.1890.

Lent by the Science Museum (Inv. No. 1970–158).

b Glass plate of solar spectrum showing the element gallium in the Sun; taken by Professor Sir W. N. Hartley FRS of Dublin.

Lent by the Science Museum (Inv. No. 1914–339).

c Three-prism spectroscope, inscribed *Troughton and Simms London 1867*.
Inv. No. 1927–806.
Science Museum photograph.

d Spectroscope mounted on the Greenwich 28in refractor, 1899.
Photograph.

e 6in and 10in photoheliograph plates taken in 1883 at Greenwich and in 1973 at Herstmonceux.

Property of RGO.

7 Eclipse and transit expeditions

a Venus in transit, December 1874; 6in square glass photoheliograph plate inscribed *1874 Honolulu*.
Ref: PM16.

b Circular glass plate, 10in diameter, with 60 exposures of Venus near Sun's limb, taken in quick succession by the Janssen method, 1874.
Ref: PM.2a.

c Various photographs of eclipse expeditions.

8 G. B. Airy – personal items

a Miniature spring-driven regulator clock, with $\frac{1}{2}$sec gridiron pendulum and 24hr dial, made for G. B. Airy by unknown maker, c.1850.

Colour plate 15

b Inscribed document conferring the Freedom of the City of London, in gold box, 1875.

c Honours and Awards:
Knight Commander of the Bath, 1872.
Grand Cross in the Imperial Order of the Rose of Brazil, 1872.
Prussian Order of Merit *(Pour Le Mérite)*, 1854.
First Class Commander of the Order of the North Star (Sweden and Norway), 1873.

Gold medals:
Pendant bearing the crowned cypher of Tsar Alexander II of Russia in diamonds; The Royal Astronomical Society, 1833 (presented for his detection of 'Long Inequality of Venus and the Earth'); The Spectacle Makers' Company.

Silver medal:
Head of Nicholas I of Russia.

Bronze medals:
For services in connection with Great Exhibition 1851; head of King Christian VIII (Royal Society of Sciences, Denmark), 1852; head of Pope Pius IX, 1854; memorial medal of Stephen Hoogendijk (celebration of centenary of Experimental Philosophical Society, Rotterdam) 1869; head of A. C. Becquerel, 1874; medal commemorating Centennial Congress of Italian Scientists, Modena, 1882; head of Christopher Hansteen, 1856; King Charles and Queen Louisa, Norway and Sweden; Isaac Newton; Keith Prize of Royal Society of Edinburgh, bearing bust of John Napier.

d 'Astronomy', lithograph cartoon of Sir George Biddell Airy KCB by 'Ape', Men of the Day No. 115, *Vanity Fair* 1875.

e Passports of G. B. Airy, 1829 and 1868.

All the above lent by the Airy family.

9 William H. T. Christie, eighth Astronomer Royal, 1881–1910

10 Sir William Christie

Oil painting, signed *Jacombe Hood, 1911*.
115.6 × 90.2cm/$45\frac{1}{2}$ × $35\frac{1}{2}$in.
Sir William is seen standing in front of the 28in refracting telescope.

Lent by E. W. H. Christie Esq.

11 Greenwich as Longitude Zero, 1884

a Airy's transit circle defining the meridian, from *The Graphic*, 8 August 1885.
Photograph.

12 Visitation Day at the Royal Observatory, 5 June 1897

In the front of Flamsteed House from left to right – R. B. Clifton, A. W. Rücker, Sir G. G. Stokes, W. D. Barber, Lord Rayleigh, J. W. L. Glaisher, G. H. Darwin, Earl of Rosse, A. A. Common, W. Huggins, F. W. Dyson (Chief Assistant).
RGO Photograph.

13 Illuminated testimonial from the staff on the occasion of Sir William Christie's retirement, 1910

Four views of the Observatory in cartouches.

Lent by E. W. H. Christie Esq.

14 Frank W. Dyson, ninth Astronomer Royal, 1910–33

15 Sir Frank Dyson

Oil painting by Ernest Moore 1932, 91.4 × 76.2cm/36 × 30in.

Lent by Mrs C. S. Christie.

16 Time and eclipses

a Shortt free-pendulum clocks.
Photograph.

b Eclipses and the 'Einstein effect'.
Photograph.

17 Harold Spencer Jones, tenth Astronomer Royal, 1933–55

18 The Royal Observatory moves to Herstmonceux

19 Harold Spencer Jones – personal items

Knight of the Order of the British Empire
1955.
Knight Bachelor 1943.
Royal Medal of the Royal Society 1943.

Gold medals:
The Royal Astronomical Society 1943; The
British Horological Institute 1948; The
Astronomical Society of the Pacific 1949;
Stoke-on-Trent Association of Engineers

1950; Indian Association for the Cultivation
of Science 1953; The Astronomical Society
of Edinburgh 1953.

Silver medals:
Quatercentenary of the death of Copernicus
1943; Polish Academy of Literature and
Science 1948; The Rittenhouse Medal of
the Rittenhouse Astronomical Society,
Philadelphia, USA 1955; Institut de
France 1955.

Bronze medals:
Janssen Prize Medal, Société Astronomique
de France 1945; University of Brussels
1947; Centenary of the discovery of the
planet Neptune 1946; Belgian
Astronomical, Meteorological and
Geophysical Society 1947; University of
Liège – En souvenir de la réunion du
Comité Exécutif de l' U.A.I. 1945.

Red case containing a) red and blue tippet,
b) silver medal and c) programme for
Séance Solennelle de Rentrée, 26
November 1949.
Box containing a) blue silk tippet
b) programme for Séance Académique
Solennelle Université de Bruxelles 24
January 1947 and c) letter from Secretary
of Académie Royale de Belgique.

Area 5

The Building and Instruments at Greenwich

1675–1950

All the telescope models in this room were made by Preview of Westerham. Scale 1:8.

1 Why Greenwich?

a Two views of Greenwich before the Civil War, by an unidentified artist.

(i) View towards London from One Tree Hill with Duke Humphrey's Tower, later the site of the Observatory, on the left.
Panel 43.2 × 77.5cm/17 × $30\frac{1}{2}$in.
Ref: 31–11.

(ii) Prospect from what is now Point Hill, with Duke Humphrey's Tower on the right.
Panel 29.4 × 92.3cm/$11\frac{1}{2}$ × $36\frac{1}{4}$in.

Lent by the London Museum.

2 Phase I – development under Flamsteed 1675–1719

a Ground plan of the Observatory buildings in 1676.

b Etchings of the Royal Observatory by Francis Place, after Robert Thacker, *c*.1676.

(i) *Vivarium Grenovicanum,* the title plate, inscribed:
R. Thacker delineavit. F. Place Sculpsit. 10.1 × 44.0cm/$4\frac{1}{4}$ × $17\frac{1}{4}$in.

Photograph by courtesy of the Master and Fellows, Magdalene College, Cambridge.

(ii) *Ichnographia Speculae Regiae Grenovici,* plan of the Observatory.
21.5 × 35.4cm/$8\frac{1}{2}$ × 14in.

(iii) *Prospectus Australis,* view from Flamsteed House roof, looking south.
20.2 × 29.8cm/8 × $11\frac{3}{4}$in.

(iv) *Prospectus Septentrionalis,* view of the Observatory looking north-east.
16.8 × 29.6cm/$6\frac{1}{2}$ × $11\frac{1}{2}$in.

(v) *Prospectus Orientalis,* view from Flamsteed House roof, looking east, with the well telescope in the foreground.
20.3 × 29.8cm/8 × $11\frac{3}{4}$in.

(vi) *Facies Speculae Septen:* view of Flamsteed House, looking south.
17.0 × 29.4cm/$6\frac{3}{4}$ × $11\frac{1}{2}$in.

(vii) *Prospectus versus Londinium,* view of Flamsteed House, looking north-west towards London.
17.0 × 59.3cm/$6\frac{3}{4}$ × $23\frac{1}{4}$in.

Lent by the British Museum.

(viii) *Puteus 100 pedum . . . ,* the well telescope; and *Partes instrumentorum . . . ,* parts of various instruments.
On one plate 63.2 × 14.6cm/25 × $5\frac{3}{4}$in.

Lent by Greenwich Local History Library.

(ix) *Prospectus Intra Cameram Stellatam,* interior of the Great Room.
21.1 × 29.5cm/$8\frac{1}{4}$ × $11\frac{1}{2}$in.

(x) *Domus Obscurata . . . ,* darkened chamber used for observing Sun spots and eclipses; and *Quadrans Muralis . . . ,* the 10ft mural quadrant.
On one plate 19.3 × 29.8cm/$7\frac{1}{2}$ × $11\frac{3}{4}$in.

(xi) *Facies Sextantis Anterior . . . ;* and *Fanis Sextantis Posterior . . . ,* two views of the 7ft sextant.
On one plate 19.3 × 29.8cm/$7\frac{1}{2}$ × $11\frac{3}{4}$in.

Except where otherwise shown, all the above etchings lent by the Society of Antiquaries of London.

c Map of Greenwich Park *c*.1676; this may be part of the Place series of etchings.
53.3 × 47.0cm/30 × $18\frac{1}{2}$in.

Lent by the Greenwich Local History Library.

d Two drawings of the Observatory 'from life' by Gasselin, 1699 and 1702.

Photographs.

(i) *veuéé de lantréé de lobservatoire a pres nature le 4 de septembre 99,* view looking west.

(ii) *veue du derriere de lobservatoire de grenuche désynay a pres nature par Gasselin en juillet 1702,* view looking north.

e Flamsteed's 7ft sextant, in use 1676–98. Model.

f Flamsteed's 7ft mural arc, in use 1679–1719. Model.

3 Phase II – development under Halley and Bradley 1720–62

a Ground plan showing the growth of the buildings 1720–62.

b Set of 3 pen-and-wash plans of the Observatory *c.*1750–60, all signed: *Iohn Evelegh delin^t.*

   (i) *Section* of Flamsteed House.
       32.0×49.5cm/$12\frac{1}{2} \times 19\frac{1}{2}$in.

   (ii) *Base Plan* of Flamsteed House and plan of the Great Room.
       32.0×49.5cm/$12\frac{1}{2} \times 19\frac{1}{2}$in.

   (iii) *General Plan* of the Observatory buildings and grounds.
       49.7×35.6cm/$19\frac{1}{2} \times 14$in.

Property of RGO.

c *A Map of Greenwich Park by John Morton, c.1780*; with inset view of Flamsteed House. Manuscript; pen and ink on parchment.
53.3×55.9cm/21×22in.
Ref: ART/7.

d Observing rooms built under Bradley; drawing, 1769, *Plate 1.*, from *Lettres Astronomiques*, by Jean Bernoulli, Berlin 1781,
RGO photograph.

e Halley's 5ft transit instrument, in use 1721–50; now in Bradley's Middle Room, ORO.
Model.

f Halley's 8ft and Bradley's 8ft mural quadrants mounted on the east and west faces of a common wall; in use 1725–1812 and 1750–1825. The original instruments are in the Quadrant Room, ORO.
Model.

g Bradley's 8ft transit instrument, in use 1750–1816; now in Bradley's Middle Room, ORO.

   (i) *Charnock's Views Vol. IV Mechanics* by John Charnock [1756–1807], folio 12. Pen-and-wash drawing. Photograph.

   (ii) Model.

h *Charnock's Views Vol. IV Mechanics* by John Charnock, folio 14. [Book] plate showing:

  left   Bradley's $12\frac{1}{2}$ft zenith sector in use at Greenwich 1747–1812. Now in the Quadrant Room, ORO.

  top   one of Maskelyne's 5ft equatorial sectors, east and west; east in use 1785–1811, west in use 1792–1837.

  right Bradley's 8ft quadrant (see 3f above).

4 Phase III – development under Maskelyne and Pond 1765–1835

a Ground plan showing the growth of the buildings 1765–1835.

b Flamsteed House, view looking west. Watercolour; on the back in Margaret Maskelyne's hand: *Royal Observatory Greenwich as it was in 1794 drawn by Mr Sinley.*
36.0×47.0cm/$14\frac{1}{4} \times 18\frac{1}{2}$in.

Lent anonymously.

c Flamsteed House, view looking south, 1801. Watercolour signed: *M. Maskelyne Delin^t.* '*O1* (daughter of the Astronomer Royal aged 14).
37.0×54.0cm/$14\frac{1}{2} \times 21\frac{1}{4}$in.

Lent anonymously.

d View from the courtyard of the observing rooms as extended under Maskelyne and Pond. Pencil drawing inscribed: *11 February 1839 E.S.* by Miss Elizabeth Smith. Bound in *Views of Greenwich Observatory.*

Property of RGO.

e Early illustration of Pond's time ball, erected 1833. Pen-and-wash drawing signed: *F. Earp.*
33.6×24.7cm/$13\frac{1}{4} \times 9\frac{3}{4}$in.

f Another early view of the time ball; origin of engraving unknown. Photograph.

g Maskelyne's 6ft mural circle, in use 1812–51; now in Pond Gallery, ORO.
Model.

h Pond's 10ft transit instrument, in use 1816–1850; now in Bradley's Transit Room, ORO.
Model.

5 Phase IV – development under Airy 1835–81

a Ground plan showing the growth of the buildings 1835–1881.

b View of the new rooms added on to the west side of Flamsteed House.
Pencil drawing dated *April 27 1838* by Miss Elizabeth Smith.
Photograph.

c The observing rooms looking north-east. Pencil drawing dated: *July 20th 1838* by Miss Elizabeth Smith.
Photograph.

d The skyline looking north from the kitchen garden, 1863. Watercolour by Miss Christabel Airy, daughter of the Astronomer Royal. $16.3 \times 28.6 \text{cm}/6\frac{1}{2} \times 11\frac{1}{4}\text{in}$.

   Property of RGO.

e The magnetic observatory on the south ground. Watercolour by Miss Christabel Airy, 1847. $26.0 \times 36.8 \text{cm}/10\frac{1}{4} \times 14\frac{1}{2}\text{in}$.

   Property of RGO.

f The main gate and the gate clock, c.1870; early wet plate photograph, c.1870.

g Airy's altazimuth instrument, in use 1847–97. The original instrument is now in Pond Gallery, ORO.

   (i) The altazimuth dome, looking north. Photograph, 1863.

   (ii) The altazimuth in use; engraving from 'A Day at the Observatory' by Edward Dunkin, in *Leisure Hour*, 1862. Photograph.

   (iii) Model.

h Airy's transit circle, in use 1851–1954. The original instrument is in the Airy Transit Circle Room, ORO.

   (i) The transit circle in use at night; engraving signed *W. B. Murray*, from the *Illustrated London News*, 1880. Photograph.

   (ii) The transit circle in use by day; engraving from *The Graphic*, 8 August 1885. Photograph.

   (iii) The transit circle viewed from the courtyard along the Prime Meridian. Photograph c.1925.

   (iv) Model.

i Airy's 12.8 inch Merz 'Great Equatorial' telescope, in use at Greenwich 1860–1950; now in Dome E, RGO.

   (i) Raising the North Pier into a vertical position by sheer legs at ground level. Pencil drawing dated *1857 Apr. 27*. $30.5 \times 24.2 \text{cm}/12 \times 9\frac{1}{2}\text{in}$.

   Property of RGO.

   (ii) The North Pier set up at ground level. Photograph 5 May 1857.

   (iii) The Drum Dome, with shutters open. Photograph 1890.

   (iv) *The Great Equatorial Telescope In The Dome, Greenwich Observatory;* coloured engraving signed *Fleming*, from Edward Walpole, *Old and New London*, Vol. VI, page 223, c.1870. $14.8 \times 14.8 \text{cm}/5\frac{3}{4} \times 5\frac{3}{4}\text{in}$.

   (v) Model.

6 **Phase V – development from Christie's time until the end of the Second World War 1881–1945**

a Ground plan showing the growth of the buildings 1881–1944.

b The magnetic pavilion in the Christie enclosure. Photograph c.1890.

c The Observatory buildings looking north from the Lassell Dome, across the Magnetic Ground, to the new 'Onion' Dome. Photograph c.1894.

d The observing domes looking from the roof of Flamsteed House towards the new altazimuth building and Thompson Dome. Photograph c.1897.

e The skyline, viewed from the north. Photograph c.1900.

f View of the New Physical Observatory, with Christie's altazimuth building in the foreground. Photograph c.1931.

g Visitors on the Meridian line. Photograph c.1931.

h The courtyard in 1933; in the left foreground are the transit pavilion and Cookson hut. Photograph 1933.

i The Onion Dome and the recently constructed statue of General Wolfe. Photograph c.1938.

j Bomb damage. Photograph 1944.

k Christie's 28 inch refracting telescope in use at Greenwich on the same mounting as the earlier 'Great Equatorial', 1893–1939. Now re-installed on the South-East equatorial building, ORO.

(i) The Onion Dome under construction.
Photograph 1893.

(ii) The refractor in use for double star
observations.
Photograph 1898.

(iii) Diagram of the 28 inch telescope;
scale 1:8.

l Christie's 13 inch astrographic telescope,
in use at Greenwich 1890–1939. Now in
Dome D, RGO.

(i) The astrographic dome.
Photograph 1925.

(ii) The astrographic telescope as
illustrated in the *Astrographic
Catalogue 1900.0 Greenwich Section*,
Edinburgh, 1904.
Photograph.

(iii) Model.

m Christie's 26 inch refracting and 30 inch
reflecting telescopes, in use at Greenwich
1898–1939, during which time they were
mounted together. The refractor is now in
Dome E, and the reflector in Dome A, RGO.

(i) The New Physical Observatory and
the Thompson Dome.
Photograph *c.*1899.

(ii) The refractor and reflector mounted
together inside the Thompson Dome.
Photograph *c.*1925.

(iii) Model.

n Dyson's 36 inch Yapp reflecting telescope,
in use at Greenwich 1934–9. Now in Dome
B, RGO.

(i) The Yapp Dome erected in the
Christie enclosure.
Photograph 1934.

(ii) The Yapp reflector mounted within the
dome.
Photograph 1934.

(iii) Model.

o Spencer Jones' reversible transit circle, in
use at Greenwich *c.*1936–9; now in the
Spencer Jones Group, RGO.

(i) The pavilion built in the Christie
enclosure to house the transit circle.
Photograph 1936.

(ii) The reversible transit circle in position,
with the shutters open.
Photograph *c.*1936.

(iii) Model.

7 Foreign visitors to Greenwich

a Heinrich Ludolf Benthems, German,
visited the Observatory in 1687.
His *Engeländischer Kirch und Schulen-Statt*,
was published in 1694.

Lent by the British Library.

b Zacharias Conrad von Uffenbach, German,
visited the Observatory in 1710.

(i) His *Merkwürdige Reisen Durch
Niedersachsen, Holland und Engellend*,
was published in 1753–4.

Lent by the British Library.

(ii) *London in 1710 from the travels of
Zacharias Conrad von Uffenbach*,
translated and edited by W. H.
Quarrel and Margaret Mare, published
Faber & Faber Ltd, 1934.
Translation of London part of (i)
above.

Lent by Greenwich Local History Library.

c Jean Bernoulli, b.Basle 1744, d. Berlin
1807, mathematician and astronomer,
visited the Observatory in 1769.
His *Lettres Astronomiques*, were published
in Berlin, 1771.

Property of RGO.

d Thomas Bugge, Danish astronomer,
visited the Observatory in 1777.
His manuscript diary for 1777.

*Lent by Det Kongelige Bibliotek,
Copenhagen.*
Ref: Ny kgl.Saml.377e, 4to.

Area 6

East Side of Great Hall

The Royal Greenwich Observatory at Herstmonceux

Recent Astronomical Research

1 Automatic measurement of photographic plates.

2 Determination of stellar positions.

3 The distances of the nearest stars.

4 Clusters and stellar evolution.

5 Radial velocities.

6 The motions of the stars in the galaxy.

7 The dynamics of globular clusters.

8 *RR Lyrae* variables.

9 Chemical abundance in stars.

10 The nature of quasars.

11 X-Ray stars and black holes.

12 Electronographic image tubes.
   a Image tube.

   Property of RGO.

Astronomy and the Earth

13 Almanacs and navigation.

14 Greenwich Mean Time and Atomic Time.

   a The Geochron time-telling device by the Kilberg Geochron Corporation of San Mateo, California. Modern.

15 The rotation of the Earth.

16 The Sun and the Earth.

Recent Astronomers Royal and Directors

17 Richard Woolley, eleventh Astronomer Royal, 1956–71.

18 Margaret Burbidge, Director, 1972–3.

19 Alan Hunter, Director, 1973.

20 Martin Ryle, twelfth Astronomer Royal, 1972.

21 **1975**

   a Model of 98in Isaac Newton Telescope (INT), Herstmonceux 1967.

   Property of RGO.

   b Model of Orbiting Astronomical Observatory (OAO–3), launched 21 August 1972.

   Lent by National Aeronautics and Space Administration, Washington DC.

22 The Anglo-Australian Telescope (AAT).

23 Astronomy with the Anglo-Australian Telescope.

24 The role of the Royal Greenwich Observatory today.

25 Looking ahead.

Colour Plates

1
John Harrison's first marine
timekeeper, 1735.
Property of the Ministry of
Defence (Navy).

2
Abraham Sharp's double
refracting telescope on an
equatorial mounting,
c. 1700.
Property of the Yorkshire
Museum York.

3
Portuguese Carracks; oil on panel, school of Patinir, c. 1525.

4
27 inch terrestrial globe by John Senex of London, c. 1740.

5
27 inch celestial globe by John Senex of London, c. 1740.

6
Cross-staff, back-staff, 24
inch Gunter's scale and 12
inch Gunter's scale; ivory
presentation set by Thomas
Tuttel of Charing Cross,
London, c. 1700. Half-hour
sandglass; turned ebony
and ivory, 17th century.

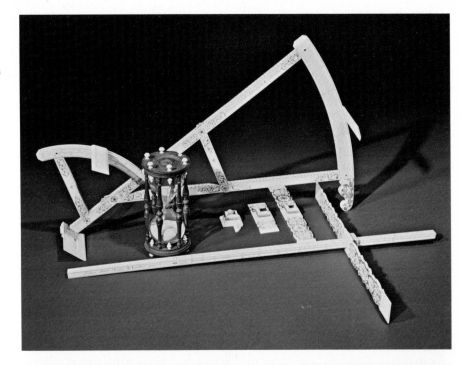

7
Universal equinoctial ring
dial; brass, inscribed H.
Bedford by Holborn
Conduit, c. 1680.

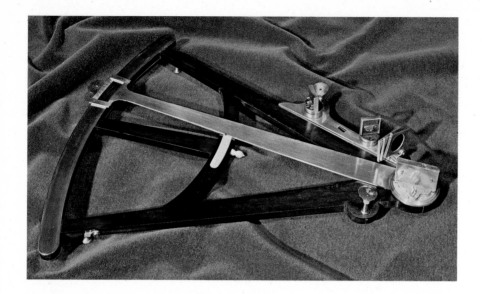

8
18 inch Hadley reflecting
quadrant; brass scale,
English, c. 1770.

9
12 inch astronomical
quadrant; brass, inscribed
John Bird London, c. 1768.
This quadrant is said to
have been taken by Captain
Cook on at least one of his
voyages between 1768 and
1780.
Property of the Science
Museum.

10
8 inch pillar frame sextant;
brass with wooden box,
signed *Ramsden London 1421*,
c. 1795.

11
24 inch Gregorian reflecting
telescope; brass, inscribed

James Short London $\frac{51}{1267} = 24, 1763$

**Property of the Whipple
Science Museum,
Cambridge.**

12
Astronomical regulator
clock on wooden tripod,
inscribed *Royal Society No. 35
John Shelton London,* 1768.
**Property of the Royal
Society.**

13
Marine chronometer No.
2241 by Charles Frodsham
of London, c. 1850.

14
Steering compass by E. J.
Dent of London, c. 1850.

15
Miniature spring-driven
regulator clock with half-
second gridiron pendulum
and twenty-four hour dial,
made for G.B. Airy by
unknown maker, c. 1850.

16
24 inch rolling parallel
ruler, brass, by Crichton of
London, c. 1820. Dividers;
brass, by Thomas Jones of
Charing Cross, London,
c. 1840.